Jasmine was always in a hurry. She heard and saw bits of things. Then she would imagine the rest.

One day, Jasmine heard Violet talk about a party. Jasmine began to imagine a birthday party of her own.

Jasmine could imagine tacking up this notice for all to see.

Jasmine would meet the King and Queen of Siam.

A friendly giant would hand Jasmine
a bunch of red roses.

Jasmine would get a flashy new automobile.

She would get service with a smile.

Then her picture would be on the
front of *Preteen Magazine*. Imagine that!